Google Home

*The Google Home Guide and
Google Home Manual with
Setup, Features, and Tips*

Matthew Adams

are for clarifying purposes only and are the owned by the owners themselves, not affiliated with this document.

Contents

Introduction

I want to thank you and congratulate you for purchasing the book, *"Google Home: The Google Home Guide and Google Home Manual with Setup, Features, and Tips."*

This book contains proven steps and strategies on how to set up and use Google Home, the latest feature product from the search engine giants. Google Home is Google's answer to Amazon Alexa. It is primarily a Wi-Fi speaker but it does a whole lot more. With Siri-like features, Google Home is your new best friend. It's a home assistant that serves you and your family and it is a control center for your smartphone. I won't go into too much detail here because that is what this book is about, teaching you how to set up Google Home and how to use it to get the best out of it.

Thanks again for purchasing this book, I hope you enjoy it!

Chapter 1

Basic Setup

So, you have your Google Home but do you really know what it is capable of or how to use it? Most likely you know a little about it so here is what you can do with Google Home:

Media:

- Play music from the most popular of the music services
- Hear the latest news from trusted sources
- Listen to your favorite podcasts
- Listen to your favorite radio station

TV and Speakers:

- Play audio on your Chromecast-enabled TV or speakers
- Stream video to your Chromecast-enabled TV
- Combine your Chromecast-enabled speakers with Google Home or Chromecast audio to play music throughout your house
- Stream audio from your mobile phone to Google Home using one of more than 100 Chromecast-enabled apps or through Android audio

Planning:

- Get the latest traffic updates and travel times for car, bicycle or walking
- Search for local places and get information about them
- Get the latest weather forecasts for any location
- Get your schedule or add events to Google calendar
- Get the latest information on an upcoming flight
- Get a "daily snippet" about your day which contains news, weather, commuter updates and reminders

Manage:

- Set timers, pause, resume, check and cancel them
- Set multiple and recurring alarms, as well as cancel, snooze or stop them
- Make a shopping list, add to it and check what is on it

Get Answers:

- Get the facts and information about whatever you want to know
- Get the latest financial new and stock prices
- Get the latest sports scores, updated and information on games
- Perform simple and complex calculations
- Translate phrases or words in any supported language
- Get the unit conversions you need
- Get nutritional information about foods or ingredients
- Get spellings and word definitions

Control Your Home:

- Control the temperature on your supported smart thermostats
- Control your lights with supported smart bulbs
- Control your devices with supported smart plugs
- If this, then that – use IFTTT to control many third-party smart devices and online services that are not integrated directly

Have Some Fun:

- Get trivia and play games
- Ask Google Home any question you want – have as much fun as you can with it

How to Set Up Google Home

Now you know what you can do with Google Home, let's get it set up.

What You Need:

- Google Home
- Latest Google Home app – Android or iOS
- Latest Google app version – Android only
- Google account – if you don't have one, set one up now for free
- iOS or Android tablet or mobile phone
- Internet connection – secure Wi-Fi is best

How to Set Google Home Up:

1. Take the power cable that came with your Google Home; plug it into the device and to a wall outlet
2. Switch on your Android mobile device and navigate to **g.co/home/setup** and install Google Home app
3. Ensure that your mobile device is connected to the Wi-Fi network that Google Home will be connected to
4. Open Google Home app on your Android device
5. Now you need to accept the privacy policy and terms of service – tap on **Accept**
6. So that Google Home can discover nearby devices and set them up, tap on **Turn on Location>Allow**
7. The app will scan for any devices nearby that are plugged in and ready to be set up. If it can't find any, make sure that you are close to the Google Home device and that it is plugged into an electric supply; tap on **Try Again**
8. If there is just one device to set up, tap on **Continue**; if there are more, locate the right one on the list and tap on it
9. Tap **Set Up** and wait for Google Home to make a connection. If it can't, get closer to the device and tap on **Try Again**
10. When the app has connected itself to the Google Home device, it will play a sound to confirm that it has connected to the correct device; tap on **Play Test Sound**
11. If you heard it, tap on **I Heard the Sound**; if not, tap on **Try Again**
12. Now choose the room where your Google Home device is – this helps in device identification when you want to cast to Google Home

13. Select which Wi-Fi network you are connecting to Google Home. If you are using Android 1 or above, you can tap on **Ok** to fetch the password for the network automatically; if not you will need to manually input the password

14. Now Google Home will try to connect to the network; if it doesn't work, check the right password has been input and tap on **Try Again**

15. When a successful connection has been made, tap on **continue**

16. Now tap on **Sign In** – this is so that Google assistant can help you with your questions and give you a personalized experience

17. Add or select the Google account that you are linking to Google Home and tap on **Sign In**

18. If you want to skip this step, tap on **Leave Setup**. However, this will mean you cannot stream music or ask Google Home any questions. If you choose not to do this, tap on **I'm Sure** on the warning window

19. If you allow Google Home to answer personalized questions, others in your home can also ask about your information. Tap on **allow**

20. Tap on **Set Location** and then on **Allow** to let Google Home access your device location

21. To allow email notifications about features, offers, apps and more, drag the slider for email notifications to the right. Leave it to the left if you don't want them and tap on **Continue**

22. If you want your cast devices to be voice controlled, you must link them up to Google Home. Find the device you are linking and tap on it; tap on **Link.**

23. Tap on **Continue** to go through a short interactive tutorial on talking to Google Home; tap **Skip** if you don't want to do this

24. Tap on **Continue** to complete the setup and you can start using your Google Home

Over the next few chapters, we are going to look more in-depth in how to set up Google Home and how to use it.

Chapter 2

Basic App Settings

Meet Google Home App

The Google Home app plays an important part in your Google Home experience. It allows you to set up Chromecast and control it. It also helps you to control Google Home devices and find new content. Chromecast support includes Chromecast audio, Chromecast ultra and TV, or speakers with built-in Chromecast.

Here are the features of Google Home app:

- **Setup** – allows you set your device up quickly and easily
- **Devices** – allows you to find available Google Home and Chromecast devices and manage them
- **Watch tab** – find movies and TV shows for already installed apps – this only works on Chromecast and Chromecast-enabled TVs.
- **Listen tab** – find audio from already installed apps
- **Discover tab** – learn how to do things with your devices and all compatible apps and find new features
- **Search** – US only – search installed apps for content

Note – the tabs you see in the Google Home app are dependent on the Google Home and Chromecast devices that are linked to your app.

Change Language Settings on Mobile Device

To set Google Home up and adjust any settings, your language must be set to English US. Here's how to it:

Android:

1. Tap on **Settings**
2. Tap on **Language & Input>Language>English (United States)**
3. Close settings

iOS:

1. Open **Settings>General>iPhone Language**
2. Select **English (US)** and tap on **Done**
3. Confirm your language selection and exit settings

Find Your Firmware Version and Settings

The firmware is the software that Google Home runs on and it will be updated on a regular basis. When new updates are available, Google Home will download them automatically OTA (Over the Air). However, to do this, Google Home must be set up properly and must be connected to an internet connection. To find your firmware version:

1. Open Google Home app

2. Tap on **Devices**

3. Look for the Google Home device card

4. On the card, tap on **Menu>Settings**

5. At the bottom, you will see **Cast Firmware Version x.xxx.xxxxx**

Google Home App

If you want to get the most out of Google Home, you need to get your settings right. To find the settings:

1. Open Google Home app

2. Tap **Menu>More Settings**

3. Check that the Google account at the top of the screen is the right one; if not, click the triangle at the right of the name and change it

Google Assistant Settings

Change your settings to customize Google assistant. These settings will take effect on all your Google assistant devices:

Music Settings:

To connect and stream your music and chose a default music service:

1. Open Google Home

2. Tap on **Menu>Music**

3. Tap the radio button on the left of the music service you want to use

4. If it is Pandora or Spotify, you must link the accounts. YouTube Red and Google play will automatically be linked when you initially set up Google Home

5. To connect an account, tap on **Link** and sign in with your credentials

6. Tap on **Unlink** if you want to remove a default service

Home Control Settings:

To enable Google Home to help control your smart home:

1. Open Google Home app
2. Tap on **Menu>Home Control**

News Settings:

To make sure you get news from your favorite trusted sources:

1. Open Google Home app
2. Tap on **Menu>More Settings>News**
3. A list of news services will appear in the order they will play in when you say "Ok, Google. Listen to news"
4. Change the list order by tapping and dragging the hash mark beside the new source and dragging it to where you want it in the list

My Day Settings:

To manage your upcoming day:

1. Open the Google Home app
2. Tap on **Menu>More Settings>My Day**

3. Choose the options to include in your My Day summary and check the box beside each one – weather, work commute, next meeting, reminders

Note: if you want to hear reminders you must set them up in the Google app, Google inbox app or Google calendar app or you can type in a query in Google search – you cannot set them by voice or through the Google Home app

Next, decide how you want My Day summary to end:

- Nothing – tap the button beside **Nothing**
- News – tap the button beside **News**

TV and Speaker Settings:

To manage devices that you can use with voice control:

1. Open Google Home app
2. Tap on **Menu>More Settings>TVs and Speakers**
3. Tap on + at the bottom of the screen to add a device
4. Check the box beside each device to add
5. Tap on **Add**
6. All your chosen devices will now show up in the TV and speakers section in the app

Shopping List Settings:

Set up your shopping lists and maintain them:

1. Open Google Home app

2. Tap on **Menu>Shopping List**

3. Now the Google keep app will open if you have it; if not, Google keep on the internet will open

4. Add items or remove them as you want

Services Settings:

This will allow you to link services:

1. Open Google Home app

2. Tap on **Menu>More Settings>Services>Link Service**

3. Sign into the service you want to set up

Google Home Device Level Settings

Changing your settings can customize Google Home device. You can manage several different things in these settings:

Device Address: this is the location address for your device. It may be the same as your home and/or work locations. This is the location used when you ask for local places, time, weather and other location-specific information unless you specify another location. To change the device address:

1. Open Google Home app

2. Tap on **Menu>More Settings**

3. Find and tap on **"Adjust Settings for This Google Home Device"**

4. Choose the device you want to configure (if you have more than one) by tapping the down arrow

5. Tap on **Device Address** and type in the address for where your device is located

Personal Results

This setting is what lets Google Home read your private information out to you, like your calendar, shopping list and flight information. Obviously, personal results must be switched on for you to hear the information:

1. Open Google Home app

2. Tap on **Menu>More Settings**

3. Tap on **"Adjust Settings for This Google Home Device"**

4. Select the device you are configuring

5. Look for the **Personal Results** section and slide the slider all the way to the right

YouTube Restricted Mode

This setting is where you can hide video or music that contains content that may not be appropriate. To manage the setting:

1. Open Google Home app

2. Tap on **Menu>More Settings>Adjust Settings for This Google Home Device**

3. Go to the section for **YouTube Restricted Mode**

4. To block content, drag the slider right and to allow content, drag the slider left

Google Account Settings

There are several settings you can manage in this section, all of which are associated with the Google Home app:

Personal Information:

There are a few settings you can change:

Your Nickname – this is name associated with your own Google account but you can change it and then teach Google assistant the new name:

1. Open Google Home app
2. Tap on **Menu>More Settings**
3. Tap on **Google Account Settings>Personal>Nickname**
4. Type the new nickname into the **"Set A Nickname"** box
5. Tap the button beside **User Default** and tap on **Play** to hear what the name sounds like
6. If you want to spell your name, tap the button beside **"Spell It Out"**
7. Type in your name using Latin characters and tap on **Play** to hear it

Home and Work Locations

These locations are synchronized to the settings for Google maps on your smartphone or tablet but you can set them manually:

1. Open Google Home app

2. Tap on **Menu>More Settings>Personal Info>Home and Work Locations**

3. Type in the addresses for your home and/or work

Changing Your Preferences:

Interests and preferences can be changed in the Google Home app:

1. Open Google Home app

2. Tap on **Menu>More Settings>Personal Info>Preferences**

3. Make the changes you want

My Activity

This is where you get to control what data Google assistant generates for you, such as latest questions you asked, the last song played or events on your calendar:

1. Open Google Home app

2. Tap on **Menu>More Settings>My Activity**

3. Now either filter your activities by date or using keywords

Devices and Device Cards

When you open the Google Home app and tap on **Menu** you will see several cards that are designed to help you get the most out of Google Home and Chromecast:

- **Found New Card** – this shows up when a new Chromecast or Google Home device is found within your area. If you want to set up a new device, tap on **Set Up** (bottom of the card). If you want to keep the device available but don't want to set it up right away, tap on **Ignore**

- **Ready to Cast** – this shows you the devices that are set and ready for use

- **Active Device** – shows the device that is active and shows you what media is playing on the device with volume and playback controls

- **Linked Devices** – shows all the devices that you linked to your account and allows you to enable customization features, like voice control

- **Ignored Devices** – this lists the available devices that you previously ignored; tap on **Ignored New Devices** and a list will show up. To set one up, tap on it.

Those are the main cards that help you with your Google Home experience. There are a few more that show up under the **Discover** tab:

- **Live cards** - helps you to configure a feature or finish a particular action on your Google Home device

- **Offer card** – shows promotional offers for apps that are installed or uninstalled. To redeem an offer, touch the card. Or tap the three dots in the top right corner to **Select, Redeem, Hide the Terms** or **View All Offers**

- **App cards** – these show you all the latest apps and games that are Chromecast enabled that you haven't yet installed. To download an app on **Open App**

- **Feature cards** – shows you tips on making the most out of Google Home

- **Collection cards** – shows a collection of apps

Notifications

If a Google Home or Chromecast device is detected as not being set up, a notification tab will appear near to the **Devices** Tab. Tap on **Devices** and locate the device card

If you are playing some kind of media from a Google Home or Chromecast device, a yellow sign will show up near to the **Devices** Tab. Tap on **Devices** and scroll down to find the appropriate card

Personalization

Inside the **Watch** and **Listen** tabs, you will find a number of TV shows, podcasts, music and other content based on media content already watched. You will also find personalized offers in the

Watch tab, also based on your interests and activity. Along with that you will see:

- **Recommended for you** - content based on your interests and activities

- **Featured/Popular/Trending** - **content** suggestions based on what other people are viewing

At the bottom of the menu, there is an **App Launcher**, which shows all your installed Chromecast-enabled apps.

Which Tabs Will You See?

What you see depends on the Google Home and Chromecast devices you have set up:

Devices	Available Tabs
Chromecast only	discover, watch
Chromecast audio only	discover, listen
Google Home only	discover
Google Home + Chromecast	discover, watch
Google Home + Chromecast audio	discover
Chromecast + Chromecast audio	discover, watch

Google Home + Chromecast + Chromecast audio   discover, watch.

Chapter 3

Google Home Features

Listening on Google Home - Music

There are plenty of ways to listen on Google Home, including music, podcasts, news and radio.

Music

Google Home supports one account per streaming device.

Basic Voice Control for All Providers:

To ask Google Home to do any of the following, start by saying, **"Hey, Google"** or **"Ok Google"** then say what you want to be done"

To do this then	Say "Hey, Google" and
Ask for a song	play ...
	play ... by ...
	play ... on ... music service
Ask for a specific artist	play ....
	play ... on... music service
Ask for a specific album	play ...

19

	play ... by ...
	play ... by ... on ... music service
Ask for music based on	play happy music
Mood, genre or activity	play classical music
	play ...(genre) on ... music service
Play suggested personalized	play some music
Content from a specific service	play ...(genre) on ... music
service	
Play music on your video, TV or	play music on my
bedroom TV	
Speaker device	play music on living room
speakers	
To pause	pause
	pause the music
Resume	resume
	continue playing
Stop	stop
	stop the music
Play the next song	next
	skip
	next song
Ask what is playing	what is playing?
	what song is playing?
	what artist is playing?
Control the volume	set volume to ...
	set volume to ...%

Advanced Voice Commands for Subscription and Free Services

If you want to hear specific artists, albums, and songs, you must buy a premium subscription to Google play music. If you do not have the premium subscription you will hear music that is similar to the songs, albums or artists that you have selected. The Google play music account will be the Google account used with Google Home and there is no way to change this.

To do this	Say "Ok Google" followed by
Play a curated list from library	play ... (name the list)
Play an album	play ...
Play the previous song	back
	previous
Skip forward	skip forward ... seconds
Shuffle playlist	shuffle
Repeat the song	play it again
	repeat the song
	play this song again
Play the playlist in a loop	repeat on/off

Important note: you cannot play purchased or uploaded music directly on Google Home but you can add it to a playlist and use the command. **"Hey, Google play ... playlist"**

You also cannot request a podcast from Google play music by saying **"Hey, Google, play my Google play podcast"**. You can, however, say **"Hey, Google, play ... (podcast name) podcast"**

Pandora Music Service

To listen to Pandora through Google Home you must first link your Pandora account. You cannot request specific artists, albums or songs, instead you will hear a selection of music similar to your selected artists, albums or songs:

To do this	Say "Hey, Google" then
Play personalized radio	play my shuffle
	play my shuffle radio
Play personalized radio based On thumbed songs*	play my thumbprint radio
Play a playlist	play classic country
	play ... (Pandora station name)
Like or dislike a Pandora song	thumbs down
	thumbs up
	I like/don't like this song

* to play thumbprint radio, you need to have several stations saved and have given several songs a thumbs up for Pandora to know what you like

YouTube Music

You will need to have a subscription to YouTube red to be able to hear specific music and to do most things on YouTube.

To do this	Say "Hey, Google" then
Like or dislike a song	I like this song
	I dislike this song
	thumbs up/thumbs down

Like or dislike the current station	save this station
	unsave this station
	follow this station
	unfollow this station
Play the previous song	back
	previous
Skip forward	skip forward ... seconds
Shuffle	shuffle
Repeat the song	play this song again
	play it again

Important note: if you say, **"Ok Google, stop casting"** or **"Ok Google, stop"** when you are playing music from YouTube, it will not stop, only pause the music. To stop it completely you must end the session manually through the Google Home app:

- Open Google Home app
- Tap on **Devices**
- Scroll down to find the card for the Google Home device being cast from
- Tap on **Stop Casting**

Spotify

You must have a premium Spotify account and you must link it with your Google Home device:

To do this	**Say "Hey, Google" then**
Play liked songs from your library	play my songs

	play my library
Like or dislike a song	I like this song/dislike this song
	thumbs up/down
Play a playlist	play
Like or dislike the current station	save this station
	unsave this station
	follow this station
	unfollow this station
Play the previous song	back
	previous
Skip forward	skip forward ... seconds
Shuffle	shuffle
Repeat the song	play it again
	play this song again
Play the playlist in a loop	repeat on/off

Setting Your Default Music Service

Your default service will be the first one used wherever it is possible:

1. Open Google Home app
2. Tap on **Menu>Music**
3. Tap the button beside the service you want to use as your default

Linking Service Functionality:

- **YouTube Red and Google Play Music -** your accounts will be linked automatically when Google Home is set up

- **Spotify/Pandora -** you will need to manually link the accounts – tap on **Link**, sign in. To unlink a service just tap on **Unlink**

Other Ways of Controlling Music

You can control your music through the Google Home app or by tapping the top of your Google Home device:

Google Home Device:

To do this	Touch the device
Play, stop or pause device once	tap on the top of the
Turn the volume up	swipe the top of the device clockwise
Turn the volume down	swipe the top of the device counterclock wise

Start your music request   press the top of the device and hold it

Google Home App

1. Ensure that your smartphone or tablet is on the same Wi-Fi network as your Google Home device
2. Tap on **Devices**
3. Find the Google Home device card

The following information will be on the card:

- Content provider
- Title of the song, station or TV episode
- Artist – if it is available
- Collection – album, playlist, radio station, show series if the information is available

You can also pause, stop, resume and control the internal speakers on the Google Home device

Chapter 4

Listening to Shows and Podcasts

On Google Home, you can listen to shows, podcast and radio on your Google Home device from any room in your home just by using your voice.

Here are a few examples on how to control shows and podcasts:

To do this	Say "Ok Google" then
Listen to a specific podcast	listen to … podcast
Continue a podcast	continue listening to … podcast
Listen to the latest episode of a podcast	listen to the latest episode of … podcast
Listen the last or next episode	previous episode

	next episode
Pause	pause the podcast
	pause
Resume	resume
	continue playing
Stop	stop
	stop the podcast
Play the next or the previous podcast	next
	skip
	next podcast
	previous
	previous episode
What's playing	what podcast is playing?
	what is playing?

Note – you can talk to Google Home assistant while you are playing a podcast or a show; when you ask it a question or say something to the assistant, the podcast or show will pause and will then carry on when you are done talking with Google assistant

Other Ways to Control Your Podcasts

From Google Home Device:

| **To do this** | **Touch the device:** |
| Play, pause or stop device once | tap on the top of the |

Turn the volume up	swipe on the top of the device clockwise
Turn the volume down	swipe on the top of the device counter clockwise
Start your request	press on the top of the device and hold

From the Google Home App

1. Ensure that your tablet or smartphone is connected to the same Wi-Fi network that your Google Home device is connected to

2. Open Google Home app

3. Tap on **Devices**

4. Locate the card for the Google Home device that is laying your show or podcast

5. Using this, you can control your podcast or show – pause it, stop, resume, and change the volume

Radio

You can listen to radio stations using your voice on your Google Home device or through a Chromecast-enabled speaker or TV.

To do this	Say "Ok Google" then
Listen to the nearest radio station by name	play ... (name of radio station)
Listen to a radio station by call-sign	play ... (call-sign)
Listen to a radio station by frequency	play ... (frequency)
Listen to a radio station by frequency in	play ... (frequency) in ... (location) A different

	location
Play radio on a Chromecast TV or speaker	play ... (statio n) on ... (device name)
Play a nearby NPR station	play NPR
Pause	pause
	pause the radio
Resume	resume
	continue playing
Stop	stop
	stop the news
Find out what station is playing	what's playing?
	what station is playing?
Control the volume	set volume to ...
	set volume to ...%

Other Ways to Control Radio

From Google Home Device:

To do this	Touch the device
Play, stop or pause device once	tap on the top of the
Turn the volume up	swipe on the top of

	the device clockwise
Turn the volume down	swipe on the top of the device counterclockwise
Start request	press the top of the device and hold

From Google Home App

1. Ensure that your mobile device is connected to the same network that your Google Home device is connected to
2. Tap on **Devices**
3. Find the card that shows the radio stations playing
4. You can control the volume or stop, resume and pause the music via the card

How to Control Restricted Content

You can restrict video or music content that contains explicit content from the music or radio services including YouTube and Google play music – this content will not be played on your Google Home device:

YouTube Music and Video

1. Open Google Home app
2. Tap on **Menu>More Settings**
3. Find the section called "**Adjust Settings for This Google Home Device**"
4. Tap on **YouTube Restricted Mode**
5. Slide the slider right to block content
6. Slide it left to unblock content

Google Play Music

1. Open the Google play music website – you cannot do this via the Google play music app
2. Click on **Settings>General**
3. Check the box beside **"Block Explicit Songs on Radio"**

This will not block premium users from playing a specific album or song that is explicit

Matthew Adams

Chapter 5

Google Home and Android Audio

Google Home and Android Audio work together to let you play your favorite podcasts, music, playlists and much more. This is called mirroring:

- You can only cast your Android Audio if you are running on Android 4.4.2 or higher

- Your Android device must not be set in "Power Saving Mode" as this will reduce how much processing power your device has and this will directly affect the Cast Audio performance

- You must turn on the microphone" permission in Settings on your Android device otherwise you cannot use the "Cast Screen/Audio" feature. If it isn't switched on, your cast Audio session will stop immediately after it tries to connect. Open **Settings>Apps>Google Play Services>Permissions.** Slide the slider for Microphone to on.

Basic Voice Commands

Here are some examples on how to talk to Google Assistant to play Android Audio. Please note that most of the control commands that you use when controlling your music are not supported with Android Audio on Google Home:

To Do This	Say "Ok Google" Then
Stop Android Audio playing	Stop
Control the volume	Turn it up
	Turn it down
	Max volume

Casting Android Audio from Your Android Device

1. Make sure your Android device is on the same Wi-Fi network that Google Home is connected to
2. Open Google Home app
3. Tap on **Menu>Cast Screen/Audio>Cast Screen>Audio**
4. Tap on your Google Home and start to play your content
5.

Stop Casting

Notification Drawer

1. Pull the notification drawer down on your Android device
2. Tap **Disconnect** on your notification bar

Google Home App

1. Open Google Home app
2. Tap on **Menu>Cast Screen/Audio>Disconnect**

Controlling Your Volume Through Voice Commands

These are the voice commands you would use to control the volume

To Do This	Say "Hey, Google"	Google Home Does
Turn up volume	Turn it up	Increases by 10%
Turn down volume	Turn it down	Decreases by 10%
Set a specific level	volume level ...	Sets to the specified level
	Volume to ...%	
Maximize the volume	Max volume	Sets to highest level (10)
Minimize the volume	Minimum volume	Sets to lowest level (1)
Change by x amount	Increase volume by ...%	Changes by the specified
	Decrease volume by ...%	amount
Hear the current level	What is the volume?	Tells you the current level

Google Home volume can be changed when you have media playing, when it is paused or when there is no media playing. In the last two cases, when you change the volume, Google Home will audibly confirm it has been done

Other Ways to Control Audio When Casting Android Audio

Google Home Device

To do this	Touch the device
Turn the volume up clockwise	Swipe on the top of the device
Turn the volume down	Swipe the top of the device counterclockwise

Note – this will not adjust the volume of any timers or alarms, only the volume

Google Home App

1. Make sure that your mobile device and Google Home device are on the same Wi-Fi network

2. Open Google Home app

3. Find the Google Home device card where you will see the media playing

4. Tap on the volume button on the card to control it – moving the slider left turns the volume down and right turns it up

Chapter 6

Using Guest Mode

Guest Mode is a neat feature that makes it easy for your family and friends to cast media from their iOS or Android device to your Google Home device without the need to connect to your Wi-Fi.

Step 1: - Set Up Guest Mode – Hosts

As the host, you can maintain the privacy of your Wi-Fi network password and allow guests to cast to your Google Home so long as they are in the same room as your Google Home device:

1. Make sure your mobile devices are connected to the same network as Google Home otherwise you cannot set up Guest Mode settings.
2. Open Google Home app
3. Tap on **Devices** – this will let you see the Google Home devices available
4. Tap the one you want to set up (if you have more than one)
5. Sign in with your Google account if you haven't already
6. Tap on **Menu>Guest Mode**
7. Move the slider to turn Guest Mode on or off

Step 2 – Connecting to Guest Mode – Guests

Guests can cast their content without the need to connect to the Wi-Fi network. However, you do need to have an internet connection but this can be your mobile data connection

1. Open any app on your device that is Chromecast-enabled
2. Tap on **Cast**
3. Choose **Nearby Device** and then follow the instructions to connect it
4. If pairing doesn't work, you are going to have to enter a 4-digit OIN number that will come from the host

This 4-digit PIN is a requirement for a Guest Mode connection to be made. When a nearby device attempts to connect, Google Home will use short and inaudible tones to transfer the PIN. If it doesn't work, the guest will need to enter the PIN manually. The host can find the PIN in one of these two places:

Device Card
1. Open Google Home app
2. Tap on **Devices**
3. Locate the card for the device that you are looking for the PIN for
4. You will find the PIN underneath the device name

Device Settings
1. Open Google Home app
2. Tap on **Devices** – this will show you the Google Home devices available. Tap on the card for your device
3. On the card, tap on **Menu>Guest Mode**
4. The PIN can be seen underneath "On"

Important Notes:

Both hosts and guests that are using Guest Mode can turn Guest Mode on or off on any mobile device that provides support for the Google Home app

Guests using Guest Mode can cast content from an Android device on 4.3+ or iOS devices on iOS 8+. Bluetooth must be enabled for casting to work under Guest Mode

Guest Mode is a feature that you opt-in to, i.e. it isn't automatically enabled. Guest Mode can be managed from any Android or iOS device using the Google Home app but be aware that, if a Factory Data Reset (FDR) is performed on the Google Home device, the guest settings will be reset.

Matthew Adams

Chapter 7

Other Ways to Play Audio on Google Home

Chromecast-Enabled Apps

Google Home is the perfect way to play music through audio apps that are optimized for your speakers. You can use your tablet or smartphone as a remote control, managing volume, playback and everything else in between.

1. Ensure that your mobile device and Google Home device are on the same Wi-Fi network.
2. Open any app that is Chromecast-enabled and tap on **Cast**
3. Select the Google Home device you are casting to
4. When the connection has been made, the Cast button will change color to tell you
5. Audio can now be cast to Google Home
6. To stop casting, tap on **Cast>Stop Casting**

Remember that some providers, like YouTube, and Spotify will require you to have a subscription. In the case of YouTube, you cannot cast from the YouTube app, only the YouTube Music app.

Basic Voice Commands for Chromecast-Enabled Apps

To do this	Say "Ok Google" then
Pause	Pause the music
	Pause
Resume	Continue playing
	Resume
Stop	Stop
	Stop the music
What's playing	What is playing?
	What artist is playing?
	What song is playing?
Control the volume	Set volume to ...
	Set volume to ...%

Chrome Browser

If you use Chrome browser to listen to your entertainment you can mirror the content from your computer screen to Google Home:

Set Up Chrome Browser Casting:

1. If you don't run Chrome already, download it. Those that are running Chrome need do no more because the browser updates automatically

2. Make sure that you are on the very latest Chrome version

3. Install Google Cast Extension. This isn't a necessary step but is recommended and will give you the Cast button on your toolbar.

Casting Content

There are a few ways to cast content to Google Home from your Chrome browser. For these options, your computer must be connected to the same Wi-Fi network as your Google Home device

Cast-Enabled Sites

1. Open any website that is Cast-enabled. Example Google Play Music
2. Click on the icon for Cast
3. Select the Google Home device that you want to cast music to

Chrome Settings Menu

1. Open the menu for Chrome settings – it is in the top right corner of the browser
2. Click on **Cast**
3. Select the Google Home device you want to cast to

Chrome Webpage

1. Open a Chrome browser tab
2. Right-click your mouse on the webpage and click on **Cast**
3. Select the Google Home device you want to cast to

Google Cast Extension

1. Click on the extension button on your browser toolbar (if you installed it)
2. Select the Google Home device you want to cast to

Basic Voice Commands to Control Audio from Chrome on Google Home

These are some examples of how to manage your Chrome music content through Google Assistant:

To do this	Say "OK Google" then
Control volume	Turn it up
	Turn it down
	Max volume
Stop Chrome content playing	Stop

Chapter 8

Linking Speakers and TVs Through Google Home App

If you have speakers or a TV that is voice supported, you can connect them to Google Home. Do make sure that you have set up Google Home and that your voice-enabled devices are on the same Wi-Fi network as Google Home:

1. Open Google Home app
2. Tap on **Menu>More Settings>TV and Speakers**
3. You will now see a list of linked devices
4. To link a new Chromecast-enabled TV or speaker, tap on the + sign at the bottom of the screen
5. Google Home app will now search for new devices that are on the same network
6. Tap on the checkbox beside the name of the device you want to link – you can check several items in one go
7. Tap on **Add** and all linked devices will show up on the TV and Speaker section of the Google Home app

Note – if you get an error message when trying to link a device, the only way to clear it is to do a Factory Data Reset on your device

– more about that later

Unlinking Devices

If you decide that you want to unlink a device:

1. Open Google Home app
2. Tap on **Devices>Linked Devices**
3. Tap on the X next to the device that you want unlinked

Name Your Google Home and Chromecast Devices

To make sure you get the proper playback on your TV or speakers, follow these naming tips:

Make sure that your Google Home and all Chromecast devices are given different names:

Good Names	Bad Names
Google Home: Living Room	Google Home: Living Room Home
Chromecast: Family Speaker Speaker	Chromecast: Living Room

If the devices have been given distinct names and you are playing your media on the Google Home rather than the TV or speaker when you make a request don't use the device name. Instead say **on speaker** or **on TV** instead.

Do make sure that your devices names are easy to pronounce and don't use special characters or emoji in the name.

Changing the Name of a Device

1. Open Google Home app
2. Tap on **Devices** – this will give you a list of Chromecast, Chromecast Audio, and Google Home devices
3. Locate the device card for the one you want to change the name for
4. On the card, tap on **Menu>Settings>Name**
5. Delete the name, type in the new one and tap on **Save**

Voice-Supported TVs and Speakers

These devices can play your audio from Google Home:

- Chromecast
- Chromecast Audio
- Devices that have Chromecast built in – must be on Android 1.21+
- Android TV devices

Basic Voice Commands for Controlling Audio to TVs or Speakers

To do this	Say "Ok Google" then
Play music on a specifically	Play … (artist) using … (music service) on … (named
named device	speakers or TV)
	Pause on … (named speaker or TV)
	Play … (music) on … (name of group)
	Stop … (name of group)

If you have a single linked video device, you do not need to name it in the command. Simply five your control command and add **on Chromecast** or **on TV**

Other Ways to Control Audio to TVs and Speakers

Google Home App
1. Ensure that your Google Home device and mobile device are on the same Wi-Fi network
2. Open the Google Home app
3. Tap on **Devices**
4. Locate the Google Home device card

On the device card, the following information is available for the current playing music:

- Content provider
- Title of song, program on a radio station or show episode
- Artist name – if available
- Collection – album, playlist, radio station, show series – if available

You can also control the remote device, pause, stop or resume the music from the card

Chapter 9

Multi-Room and Group Playback

You can group together any combination of Chromecast Audio, Chromecast-enabled speakers or Google Home to get quality music throughout your entire house. All your favorite audio and music from any Chromecast-enabled app will be available to stream instantly:

Create and Manage Your Audio Groups

Step 1: Create Audio Group

1. Open Google Home app
2. Tap on **Devices** – this will show you a list of all the audio devices available (must be on the same Wi-Fi network as Google Home
3. Select that audio device you want to put in a group and tap on **Menu>Create Group**
4. The default name of a group is Home Group but you can change it – simply delete the name and type a new one in
5. Now a list of all available audio devices will be available
6. To remove a device or add one to the group, check the box beside it to indicate that it will be associated with the group – you must select at least two for a group to be created.

7. Tap on **Save**

When you have added the audio devices to your group. A group device card will show up in **Devices**. This may take a few seconds to appear. If you want to see which speakers have been added to the group, tap on **Speakers in Group.**

1. On the new group card, you will see a ribbon that gives you the option to Enable Voice Control and More". To confirm the link to the group, tap on this ribbon. You can also do this by opening **TV and Speakers** and tapping on the Assistant settings
2. To verify the link and that voice has been enabled, tap on the overflow menu on the card and click **Linked Accounts**

Step 2: Voice Enable Audio Group
1. Open Google Home app
2. Tap on **Devices** and find the group card
3. Tap on the blue banner on the card that says **Enable Voice Control and More** and then tap on **I'm In**

Edit Existing Group
1. Open Google Home app
2. Tap on **Devices** and you will see all your audio devices
3. Select the group card to be edited
4. Tap on **Menu>Edit Group** – this will show the group name and all the speakers in it

5. tap the box beside a device to add or remove it – at least two speakers must be left for a group to be maintained

6. Tap on **Save**

7. A message will appear at the bottom of the device confirming the changes have been saved

Note: If you are editing a group that is currently playing music, playback will be stopped and you won't hear anything on the speakers within the group. You must go back to the app that is cast-enabled to start casting the audio again

Delete a Group

1. Open Google Home app

2. Tap on **Devices** – this will show all the available audio devices

3. Locate the card for the group you want to delete

4. Tap **Device>Delete Group> Delete** on the card

If you are attempting to delete a device that is casting, playback will stop.

When the group has been deleted, a message will appear at the bottom of the device and the group card will show up in the **Devices** menu

Use Voice to Control Audio Group

Playing an audio group is similar to playing on a remote individual audio device except, instead of giving a device name, you will give the group name instead:

To do this:	Say "Ok Google" Then
Play music with a group name	Play Rock on ... (name of group)
Control music using basic commands	Pause ...
	Play ...
	Stop ...
	Resume ...
	Play next song ... (group name)
Control the volume	Set volume to ...
	Set volume to ...%

Note: in group playback, the volume commands are only going to change the volume on the Google Home device unless you add the group name to the command

Other Ways to Control Your Audio Groups

Google Home Device

To do this	Touch the device
Play, stop or pause (the whole group)	Tap on the top of the device once
Turn the volume up	Swipe the top of the device clockwise
Turn the volume down	Swipe the top of the device counterclockwise
Start your request	Press the top of the device and hold it

Note: adjusting the volume will only affect the Google Home device, not the other group devices

Google Home App

1. Ensure that the mobile device and Google Home device are connected to the same Wi-Fi network
2. Open Google Home app
3. Tap on **Devices** to see a list of all your Google Home devices (if you have more than one)
4. Select the device card for your Google Home device

The card will show you the following information if it is available:

- Content provider
- Title of song, radio station or show episode
- Artist
- Collection – radio station. Album playlist, show series

You can also use the device card to resume, pause stop and control members of the group

Matthew Adams

Chapter 10

View Photos

Another great feature of Google Home is being able to see your personal photos, that are stored on your Google Photo library, on any Chromecast TV using voice control. To do this your TV must have Chromecast built in or it must be Chromecast enabled and linked to your Google Home.

To see your photos on your TV through Google Home:

Step 1: Link your TV

Both your Google Home device and your TV must be on the same Wi-Fi network

1. Open Google Home app
2. Tap on **Menu>More Settings>TVs and Speakers.** A list of your linked devices will appear
3. To link a new TV, tap on the + sign at the bottom of your screen
4. Google Home app will now search for all supported TVs on your Wi-Fi network
5. To add a new device from the list that appears check the box beside the device – you can add as many as you like in one go

6. All your linked devices will now show up in the TVs and Speakers menu in the app

Note: If you are faced with an error message while trying to link your device, you will need to carry out a Factory Data Reset on your Chromecast device

Step 2: Enable Personal Results

1. Open Google Home app
2. Tap on **Menu>More Settings**
3. Look for **Adjust Settings for this Google Home Device**
4. Select the Google Home device you want to make changes to by clicking the Down arrow – if you only have the one Google Home device, you won't see an arrow
5. Go to **Personal Results** and slide the slider right to turn it on

Step 3: Allow Photos to Display on TV

1. Open Google Home app
2. Tap on **Menu>More Settings>Videos and Photos**
3. Go to the Photos section and move the slider to the right

The only photos displayed will be the ones that are associated with the Google account that you set up Google Home with

Basic Voice Commands for Viewing Slideshows

To do this   **Say "Ok Google" then**

View photos of specific people	Show photographs of ... (name) on ... (device)
View photos of places	Show photographs of ... (place name) on ... (device)
View photos by dates	Show photographs from ...(date) on ... (device)
View photos of things	Show photographs of ... (thing) on ... (device)
View photos in a specific album	Show photographs of ... (album name) on ... (device)

For these to work, your photos must have been backed up through Google Photos and organized

Basic Voice Commands for Controlling Slideshows

When you have used one of the commands to start your slideshow, the Google Photos logo will appear on your TV screen. Once the slideshow has opened and started, use on of the following commands to control it. After issuing a command, conformation of it will appear on your screen in the top left corner:

To do this	**Say "Ok Google" then**
View the next photo	Next photo on ... (device name)

	Next slide on … (device name
View the previous photo	Previous photo on … (device name)
	Previous slide on … (device name)
Pause the slideshow	Pause photos on … (device name)
	Pause slideshow on … (device name)
Resume the slideshow	Resume photos on … (device name)
	Resume slideshow on … (device name)
Stop the slideshow	Stop photos on … (device name)
	Stop slideshow on … (device name)

Note: any low quality and duplicate photos will be removed from the slideshow and the most recent photos will be shown first.

Stop Google Photos Displaying on Your TV

When Personal Results in enabled, Google Photos are shown on linked TVs by default. To turn this off:

1. Open Google Home app
2. Tap on **Menu>More Settings>Videos and Photos**
3. Go to the Photos sections and move the slider left

Chapter 11

Controlling Your Home with Google Home

One of the best features of Google Home is the ability to control your home using your voice. Here's how to do it

Connection and Control – Lights

You will need to use specific lightbulbs, either Phillips Hue or SmartThings.

1. Ensure that your mobile device is connected to the same Wi-Fi network as your Google Home
2. You must be signed into the same Google Account on your mobile device as you used when you set up Google Home

Phillips Hue Bulbs

1. You will need to use the Phillips Hue app to set up the Phillips Hue lights and bridge
2. The bridge must be connected to the same Wi-Fi network as Google Home

SmartThings

1. You must use the SmartThings app to set up your lightbulbs

Connect Your Bulbs

2. Add Lights

3. Open Google Home app

4. Tap on **Menu>Home Control**

5. Tap on the **Devices** tab and you will see all the lights that are connected

6. Tap on the + sign at the bottom of the screen to add a new light

7. Choose the name of your bulb provider – SmartThings or Phillips Hue

Phillips Hue

1. Tap on **Phillips Hue>Pair**

2. On the Phillips Hue bridge, there is a "link" button – press it

3. Tap on **Assign Rooms**

4. Wait for the pairing to complete and then you can assign lights to specific rooms (see below) or tap on **Done.**

SmartThings

1. Tap **SmartThings**

2. Log in using your username and password

3. You can now see what information is going to be used when you connect your SmartThings lights to Google Home

4. Choose the lights that are going to connect to Google Home

5. Tap on **Authorize**

6. Assign lights to specific rooms – see below

There is no limit on how many devices can be connected to Google Home

Renaming or Setting Nicknames for Lights

You can choose the name that you gave your lights in Hue or in SmartThings or you can change it.

1. Open Google Home app
2. Tap on **Menu>Devices**
3. You will see a list of the lights you have connected to Google Home
4. Tap on the lights that you want to set nicknames for
5. Tap on **Set a Nickname**
6. Type the name and tap on **OK>Device Details**
7. Your light will now be listed with its new name

Nicknames are just another way of referencing your lights within the Google Home app and you will NOT see these new names in the SmartThings or Phillips Hue apps

Change the Room Where a Light is Located

1. Open Google Home app
2. Tap on **Menu>Devices** and you will see a list of your lights
3. Pick the lights that you want to move to another room
4. Tap on **Room**
5. Tap the button beside the room that you want to move the lights to

6. If you want to add a room that is not listed, tap on **+Custom Room**

7. Tap the Back arrow

8. You will now see the lights in their new rooms

You can only add a light to one room.

Create a Room

1. Open the **Rooms** tab and you will see the list of lights you connected

2. Add a light by tapping the + at the bottom of the screen

3. Choose a room by tapping on the button beside the room name

4. Tap on **Next**

5. If a room is not listed, tap on + **Custom Room**

6. Type in the name of the room and tap on **OK**

7. You will now see a list of the linked lights; check the box beside the light that you want to add to the room

8. Tap on **Done**

Rename a Room

1. In the **Rooms** tab, you can see a list of the lights assigned to each room

2. Tap on the room name that you want to change

3. Type the new name and tap on **Ok>Done**

4. The lights will now be shown in their new room

Move Lights to a Room

1. In the **Rooms** tab is a list of the rooms and the lights assigned to them

2. Tap on the room you want to change

3. A list of the lights in that room will appear

4. Tap on **Add Devices** and then check the box beside the light that you want to move into the room; if you want all lights in that room, tap on **Select All**

5. Tap on **Done** and the lights will be shown in their new room

Move Lights Out of a Room

1. In the **Rooms** tab is a list of rooms and the lights assigned to them

2. Tap on the name of the room you want to change

3. Select the light/s you want to move and tap on **Move**

4. Tap on the button beside the room you want to move the light to

5. If a room is not listed, tap on **+ Custom Room**

6. Name the room and tap on **Ok**

7. Tap on **Done** and you will see the light has been removed from the room and plead in a new one

Delete a Room

Before you can delete a room, it must be empty so follow the above steps to move the lights out

1. In the **Rooms** tab is a list of the rooms with the lights assigned to them

2. Tap on the name of a room you want to delete

3. Tap the Trash can you see at the top

4. Tap on **Ok** to confirm that this is the room to delete

Checking for Lights

SmartThings Only

1. Open Google Home app

2. Tap on **Menu>Devices**

3. A list of rooms and the lights assigned to them will appear

4. Tap on **More>Manage Accounts**

5. Tap the account that you want to check for lights

6. Tap on **Check for New Devices**

7. Choose a location from the dropdown menu and tap on **Authorize**

8. The number of lights that are added will show up at the bottom of the screen

9. Follow the steps above to assign lights to a room

Unlink a Light

There are two ways to do this:

Google Home App – Recommended Way

1. Open Google Home app

2. Tap on **Menu>Devices**

3. A list of your rooms and their lights will appear

4. Tap on **More>Manage Accounts**

5. Tap the account that you want to unlink

6. Tap on **Unlink Account>Unlink**

7. A confirmation will appear at the bottom of the screen

SmartThings or Phillips Hue App

- If you have used the app to unlink your lights you must also follow the above steps to unlink them from Google Home

Using your Voice to Control Your Lights

You can adjust your lightbulbs with voice control; here are some examples of what to say:

To do this	Say "Ok Google" then
Turn a light on or off	Turn on ... (name of light)
	Turn off ... (name of light)
Dim the light	Dim ... (name of light)
Brighten the light	Brighten ... (name of light)
Set a light to a specific percentage	Set ... (name of light) to ...%
Dim or brighten by a percentage	Dim ... (name of light) by ... %
	Brighten ... (name of light) by ...%
Change the color of a light	Turn ... (name of light) ...(color)
Turn all lights in a room on/off	Turn off lights in ... (name of room)
	Turn on lights in ... (name of room)
Turn all lights on or off	Turn on all lights
	Turn off all lights

Mange Thermostats Using Google Assistant

You can now use Google Assistant and Google Home to control your home temperature, flicking between cooling and heating and much more besides. You need to have a Nest Thermostat or those that are connected to SmartThings:

1. Ensure that your mobile device is on the same Wi-Fi network your Google Home is connected to

2. You must sign into the Google account that was used to set up your Google Home

Manage Thermostats

Adding a Thermostat

1. Open Google Home app
2. Tap on **Menu>Home Control>Devices**
3. A list of the thermostats that are connected to Google Home will show up
4. To add a new Nest thermostat, tap the + sign at the bottom of the screen
5. Tap on the provider name – SmartThings or Nest

Nest

1. Tap on **Nest;** sign in if asked to
2. You will now be able to read about the information that is going to be used to connect your thermostat
3. Tap on **Accept>Continue**
4. Select the Nest Home that you are connecting to Google Home (if you have more than one)

5. Tap on **Got it** and you are now ready to assign your thermostats to specific rooms (see below)

SmartThings

1. Tap on **SmartThings**
2. Log in using your username and password
3. You can now read what information is going to be used to connect your thermostats
4. Choose the thermostats you want to connect to Google Home
5. Tap on **Authorize**
6. You can now assign your thermostats to specific rooms – see below

There is no limit to the number of devices that can be connected to Google Home

Setting Thermostat Names

1. Open Google Home app
2. Tap on **Menu>Devices**
3. A list of thermostats connected to Google Home will appear – tap on the one that you want to set a name for
4. Tap on **Set a Nickname**
5. Type in the name and tap on **OK>Device Details**
6. Your thermostat will now be shown with the new name

These names are just another reference to the thermostats in Google Home app and will not be shown in the SmartThings or Nest apps

Create a New Room

1. In the **Rooms** tab is a list of the rooms and the thermostats assigned to them
2. Tap the + sign at the bottom of the screen
3. Choose a room by tapping on the button next to the name and tap on **Next**
4. If you want to add a new room, tap on + **Custom Room**
5. Type the room name in and tap on **Ok**
6. A list of the linked thermostats will appear; to add a new thermostat to the room, tap the box beside the name and tap on **Done**

You can only add a thermostat to one room

Renaming a Room

1. In the **Rooms** tab is a list of the rooms and the thermostats that are assigned to each one
2. Tap on the name of the room that you want to change
3. Type the new name in and tap on **Ok>Done**
4. The thermostats will now be listed under their new names

Nest Thermostats

If you add a nickname to a thermostat within the Google app it will not be reflected in the Nest app. However, if you give a new name to your thermostat in the Nest app, it will update the name in the Google Home app

Assign or Move the Thermostat to a Room

1. In the **Rooms** tab is a list of the rooms with the thermostats assigned to them
2. Tap on the name of the room that you want to make a change to
3. A list of thermostats for that room will appear
4. Tap on **Add Device** and check the box beside the thermostats you want moved into that room. If you want to move them all, tap on **Select All**
5. Tap on **Done** and an updated list will appear

Move a Thermostat Out of a Room

1. Go into the **Rooms** tab
2. Tap on the name of the room you want to change
3. Pick the thermostat to move and tap on **Move**
4. Tap the button beside the room you want to move the thermostat to

5. If you want to add a new room tap on + **Custom Room**

6. Type the room name and tap on **Ok**

7. Tap **Done** and the thermostat will be moved out of the room

Deleting a Room

To delete a room, it must be empty first

1. In the **Rooms** tab

2. Tap on the name of the room you want to delete

3. Tap on the trash can and tap on **Ok** to confirm

Checking for Thermostats – SmartThings Only

1. Open Google Home app

2. Tap on **Menu>Devices**

3. Tap on **More>Manage Accounts**

4. Choose the account that you want checked for new thermostats

5. Tap on **Check for New Devices** and choose a location form the dropdown menu

6. Tap on **Authorize** and the number of added thermostats will show up at the bottom of the screen

7. See the steps above for assigning them to rooms

Unlinking Thermostats

You can do this in two ways:

Google Home App – Recommended Method

1. Open Google Home app
2. Tap on **Menu>Devices**
3. Tap on **More>Manage Accounts**
4. Tap on the name of the account that you want to unlink
5. Confirmation will appear at the bottom of the screen

SmartThings or Nest Apps

- If you already unlinked the thermostat through the Nest or SmartThings apps, you need to follow the above steps to unlink it in the Google Home app

Product Specific Information

Nest Thermostat

When you use Nest Thermostats with Google Home you need to be aware of these things:

- Google Home supports the use of 1st, 2nd and 3rd Generation Nest Thermostats

- Google Home will accept up to 20 voice commands an hour for your Nest thermostat – this is to help in the preservation of the battery on the Nest thermostat

- Only one Nest Home can be supported by each Google account. If there are more than one home in your Nest, choose one

- All the thermostats that are in a Nest home will be automatically imported to Google Home. You will not be able to choose specific thermostats to import and you will not be able to remove individual ones after they have been imported to Google Home

- The only person that can authorize the integration between Nest Thermostat and Google Home is the Nest home owner, not those who are sharing access through a Family account

- The location name or label is changed in the Nest app or on the thermostat itself, it will be updated in Google Home but it won't work the other way around.

- If your Nest thermostat is on Emergency heat or there is an Emergency Shutoff, Google Home cannot control the thermostat

- Structure status cannot be set to Eco, Home or Away

- If you try to set your thermostat to Eco it will register as being switched Off

- At the time of writing, Nest Indoor/Outdoor Cams and Nest Protect are NOT supported on Google Home

Voice Commands to Control Thermostats

Here as some examples of how to talk to Google Assistant to control your thermostats

To do this	Say "Ok Google" then
Adjust temperature	Make it cooler/warmer
	Lower/raise the temperature by ... degrees
	Set the temperature to ... degrees
Switch between heating and cooling	Turn on heat/cooling
	Set thermostat to cooling/heat
	Turn the thermostat to heat-cool
Set temperature and mode	Set air conditioning to ...degrees
	Set heat to ... degrees

Turn the thermostat off	Turn off thermostat - to put the thermostat back on you will have to specify which mode you want – heating or cooling.
Set the thermostats with the room name	Set ... (name of room) thermostat to ... degrees
Hear the ambient temperature	what is the temperature inside?
Hear what temperature thermostat	what is the thermostat set to?

is set to

Control Switches and Plugs

You can use voice control to control connected plugs and switches that are connected to SmartThings

1. Ensure that your Google Home and mobile devices are connected to the same Wi-Fi network
2. Ensure that you sign in using the same Google account as you set your Google Home up with

Before you can set up your SmartThings-compatible switches or plugs you must first set them up on the SmartThings App. When you have done that, you can move on to the following.

Add Switches or Plugs

1. Open Google Home app
2. Tap on **Menu>Home Control>Devices**
3. A list of the connected switches and plugs will appear
4. To add a switch or a plug, tap on the + at the bottom of the screen
5. Tap on **SmartThings**
6. Sign in using your username and password and tap on **Log In**
7. See the information that is going to be used when you connect a switch or a plug to Google Home
8. Choose the switch or plug you are adding and tap on **Authorize**
9. You are now ready to assign your devices to rooms – see below

There is no limit to how many you can link to your Google Home

Setting Nicknames

You can either use the name that you assigned to each plug or switch in the SmartThings app or you can put new nicknames in using Google Home

1. Open Google Home app
2. Tap on **Menu>Devices** and a list of the connected plugs and switches will appear
3. Tap on the one you want to give a nickname to and tap on **Set a Nickname**
4. Type the name and tap on **Ok>Device Details**
5. The new name will show up in the list

These nicknames are only another way of referencing the plug or switch in Google Home app and will not show up in the SmartThings app.

Change the Rooms Where Your Switches or Plugs Are Located

1. In the **Devices** tab, you will see a list of the connected plugs and switches
2. Tap on the ones that you want to be moved into a new room
3. Tap on **Room** and tap the button beside the room that you want it/them moved to
4. To add a new room, tap on + **Custom Room**

5. Tap on the back arrow and you will see the switch or plug assigned in its new rooms

You can only add a plug or a switch to one room

Move a Plug or Switch Out of a Room

1. In the **Rooms** tab, there will be a list of the rooms with the switches and plugs assigned to them
2. Tap on the name of the room that you want to change
3. Pick the switch or plug that you want to move and tap on **Move**
4. Tap on the button beside the room that you want it moved to
5. If you want to add it to a room not listed, tap on + **Custom Room**
6. Type the room name in and tap on **Ok**
7. Tap on **Done** and the plug or switch will be gone from the room

Deleting a room

Before you can delete a room, you must first move all plugs and switches out of it

1. In the **Rooms** tab is a list of all the rooms with the plugs and switches assigned to them

2. Tap on the room that you want to delete and tap on the trash can

3. Tap on **Ok** to confirm the room you want to delete

Checking for Switches or Plugs

1. In the **Devices** tab is a list of the rooms and the switches and plugs that are assigned to them

2. Tap on **More>Manage Accounts**

3. Select the account that you want to check and tap on **Check for New Devices**

4. Choose the location from the drop down menu and tap on **Authorize**

5. The bottom of the screen will show you how many switches and plugs have been added

Unlinking Switches and Plugs

You can do this in two ways:

Google Home App – Recommended Method

1. In the **Devices** tab is a list of all the rooms and the plugs and switches assigned to them

2. Tap on **More>Manage Accounts**

3. Choose the account you want unlinked

4. Tap on **Unlink Account>Unlike**

5. The confirmation will be at the bottom of the screen

SmartThings App

- If you unlink an account through the SmartThings app you will still need to follow the above steps to unlink it from Google Home

Voice Commands to Control Your Switches and Plugs

These are some of the ways that you can talk to Google Assistant to control your switches and plugs:

To do this	Say "Ok Google" then
Turn the plugs on or off	Turn on/off ... (name of plug)
Turn the switches on or off (Name of switch)	Turn on/off ...

Matthew Adams

Chapter 12

Managing Alarms, Timers, and Lists

Setting up alarms, timers and lists on Google Home are easy:

Voice Control for Setting and Managing Alarms:

To do this	Say "Hey, Google"	Google Home does
Set a new alarm	Set alarm for ...am/pm tomorrow	Confirms alarm time
Set alarm with name	Set alarm for ...am/pm named ... (give it a name)	Confirms the alarm has been set
Set a recurring alarm	Set alarm for ... am/pm every day of the week	Confirms recurring time of the alarm
Ask for information about an existing alarm	When is the alarm set for? When is my alarm on ...?	Tells you what your alarms are
Cancel an alarm	Cancel my alarm When is my alarm? Cancel it	Cancels the alarm

Stop an alarm ringing	Stop	Stops alarm
Snooze	Snooze	Snoozes alarm for the
	Snooze for ... minutes	specified time

Control the Volume of the Alarm

To do this	Say "Hey, Google" then	What Google does
Increase the volume by 10%	Louder	Increases volume
	Turn it up	
Decreases the volume 10%	Softer	Decreases volume by
	Turn it down	
Set a specific volume specified	Volume level ...	Sets the volume to a
	Volume to ...%	level
Maximize the volume	Max volume	Sets volume to 10
Minimize the volume	Minimum volume	Sets volume to 1
Change volume by x volume by the	Increase volume by ...%	Changes the
	Decrease volume by ...%	amount specified
Hear the current volume current level	What is the volume?	Tells you the

Control Alarm Volume by Touch

To do this	Touch the device
Turn the volume up	Swipe the top of the device

84

clockwise

Turn the volume down   Swipe the top of the device counterclockwise

Note: alarm sounds can only be played from the Google Home device internal speaker and not from any other Cast device that may be associated with your Google Home device. You also cannot change the default alarm tone

These voice controls can also be used for timers

Create Shopping Lists

Make sure that Personal Results is enabled on your device – you cannot create or manage shopping lists without this. Also, ensure that your mobile device is on the same network as Google Home and that you sign into the same Google account as you used when you set up Google Home

Voice Control for Managing Shopping Lists

To do this   **Say "Ok Google" then**

Add items to an existing shopping list   Add ... (item) to my shopping list

Add multiple items   Add ... and ... to my list

Ask what is on your list   What is on my shopping list?

Manage Lists with Google Home App

- Open Google Home app
- Tap on **Menu>Shopping List**

- Now the Google Keep app will open if you have it installed; if not, the Google Keep website will open
- Add and remove items from your list

Chapter 13

Getting Information, Answers, and More

Google Home is able to help you get definitions and answer many questions that you might have, including calculations, nutritional information, conversions and much more besides. Ask Google Home anything you like. Just say "Hey, Google" and then your questions. Some examples would be

- "Hey, Google, what is the capital of France?"
- "Hey, Google, how far away is the sun/moon?"
- "Hey, Google, how many Oscars does Meryl Streep have?"
- "Hey, Google, what is the smallest State in America?"

Google Home is also able to compute complicated calculations, such as:

- "Hey, Google, what is 574 times 21?"
- "Hey, Google, what is 22% of 754?"
- "Hey, Google, what is the 5th root of 972?"

You can use Google Assistant as a translation tool:

- "Hey, Google, how do I say Good Morning in Chinese?"
- "Hey, Google, how do I ask for a cup of coffee in Norwegian?"

And for up to the minute currency conversions:

- "Hey, Google, how many dollars to a Euro?"
- "Hey, Google, what is a dollar worth in Mexico?"

For unit conversions:

- "Hey, Google, how many pounds are there in 10 kilograms?"
- "Hey, Google, how many liters are there in 5 gallons?"

For nutritional information:

- "Hey, Google, how many grams of fiber is in 1 kilogram of kale?"
- "Hey, Google, how much sugar is there in one can of cola?"
- "Hey, Google, how many calories are there in a Snickers bar?"

As a dictionary to get spellings and definitions:

- "Hey, Google, how do I spell disparate?"
- "Hey, Google, what is the definition of proxies?"
- "Hey, Google, what does circumspect mean?"

You can use it to find out what the time is in different locations:

- "Hey, Google, what is the time now?"
- "Hey, Google, what is the time in ... (location)?"

To get some information on holidays:

- "Hey, Google, when is ... (name of holiday)?"
- "Hey, Google, how many days is it until ... (name of holiday)?"

And to find out exactly what Google Home can do for you, just ask "Hey, Google, what can you do?"

Get Your Sports Information

With Google Home, you can keep right up to date with the latest information on your favorite sports team, where they are playing, latest scores etc. Here are some of the ways you can talk to Google Assistant to get the latest Sports updates:

To do this	Say "Ok Google" then	Google Home does this
Hear the latest score the game	What was the score for the ... game?	Gives you score
Hear how a team is doing you the score of the	How are ... doing?	Tells

	current game and their win/loss record	
Hear how a team is doing in the league and their record	What is ... standing of ...? What is their record?	Tells you the standing of the team and their record
Hear the next game	When is ... next playing?	Tells you the date, location time, who they are playing
Hear the team or league schedule	Who are ... playing this week? Who's playing in the ... today? (name of league)	Tells you the team or league schedule
Hear if team lost/won?	Did ... win or lose?	Tells you whether the team won or lost
Hear the league standings	... (league) standings	The team ranked first

Supported Leagues and Sports

Not all sports or leagues are supported at the time of writing:

Sport	Leagues
Baseball	KBO (Korea Baseball Organization), MLB, Nippon Professional Baseball
Basketball	WNBA, NBA, Men's Division I NCAA
American Football	NCAA Division I FBS, NFL

Canadian Football   CFL (Canadian Football League)

Hockey   Kontinental Hockey League, NHL

Cricket   Indian Premier League, International Cricket Council

Tennis   WTA, ATP

Soccer   Global tournaments and leagues

Rugby Union   Heineken Cup, English Premiership, Currie, ITM Cup

Rugby League   National Rugby League, Super League

Motor Racing   IndyCar, MotoGP, Formula 1

Golf   LPGA, PGA

Other sports and leagues will be given support soon

Stock Market Updates

Google Home can also help you with information on stocks and on your portfolio as well as helping you stay up to date with the economy and the global market:

To do this	Say "Ok Google" then	Google Home does
Hear about individual Stock prices and the	What is ... trading at?	Tells you the current price of the specific stock and the gains or losses for the day
Hear about an index	What is ... trading at?	Tells you the current index price and the gains/losses
Hear about the health	How are the markets?	Summarizes

major indices

o

f   the

market

s

Hear about stock prices:

Before market opens   What was the ... premarket price?

Gives stock price

At opening   What did ... open at?   information   for

stock

At close   What did ... close at?

After hours   What is the after hour

price of ...?

Hear a company   What is the market cap of ...? Tells you the

market cap

m

a

r

k

e

t

c

a

p

o

f

t
h
e

s
p
e
c
i
f
i
e
d

c
o
m
p
a
n
y

Have Some Fun

It isn't all serious, you can also have a lot of fun with Google Home, get jokes, trivia, and games. Just say, "Hey, Google" then ask away!

Try out these examples and see what happens:

- "Hey, Google, play Lucky Trivia"
- "Hey, Google, play Mad Libs"
- Hey, Google, play Crystal Ball"

For fun, you can ask Google Home to entertain you, just by saying, "Hey, Google, entertain me" or ask "Hey, Google, what are your Easter eggs?"

Other ways to play include saying things like:

- "Hey, Google, sing me a song"
- "Hey, Google, do you know the muffin man?"
- "Hey, Google, is your refrigerator running?"

Chapter 14

Google Home and Your Privacy

Data Security and Privacy

Google takes your privacy seriously and allows you to control the information that is stored and shared with Google. Here is how Google collects and protect data:

Data Collection

You share the information you want about you and your personal preferences in the Google Home settings and through Google Assistant. The data that Google collects is only that designed to make your device smarter, faster and give more relevant results. Google Home only uses the information that you give permission for it to access in order to give you better answers through Google Assistant. Third party services may share information with Google Home as per their own privacy policies when you use those services through Google Home

If you agree to allow it, Google Home can access your search and location history to provide better answers. You will be asked for your address during the setup but, if you do not give it, Google

Home will guess based on your IP address.

Google will only use personal information that you have shared with it, along with your Google History in order to provide better customization. You may delete your history but this will limit the personalization of features. You can also view your history in My Activity through the setup app and at myactivity.google.com. You control your data at all times.

Data Use

Google uses data to speed up their services and make them more useful by trying to provide you with more relevant search results and better traffic updates. It is also used to provide protection from phishing, malware, and any other suspicious activity. Finally, data is used to show you relevant ads and to keep their services free for everyone to use.

Data Storage

Google stores your information on its servers within its own data centers. Conversation history with Google Assistant and Google Home is saved until you delete it. To see what Google Home has heard or what you have asked Google Home, you can go to Assistant History in My Activity or go to the setup app and tap on the My Activity link where you can delete your history if you choose. To change the information that Google has access to, go to your settings where you can edit your personal information.

Data Security

Your security is first for Google because keeping your data secure means keeping it private. Google Services are kept protected by a highly advanced security infrastructure and all conversations within Google Home are encrypted.

Data Deletion

If you uninstall Google Home, lose your mobile device or delete your Google Home account, you can go to myactivity.com and delete all your information if you want. Anything you delete is permanently gone; Google does not save a copy of your personal information. The only thing they may keep is service-related information, such as the Google products you have used, in order to improve services, prevent spam and prevent abuse.

Privacy

While Google Home may "listen" to conversations it is only listening for hot words. If those hot words are not heard, the conversation snippets are deleted permanently. When you say "Ok Google", the LEDs on the top of your Google Home device light up to indicate that Google Home is recording you and sending the recording to Google. Those recordings can be deleted at any time through My Activity

Anyone who is near to your Google Home device is able to request information from it and, if you have provided access via your Google Calendar, Gmail or any other personal information, those people can ask about the information from Google Home.

Google does not sell your information although, in certain circumstances, they may share information with third parties. If you use Google Home to request services from businesses, Google sends your information to the service to complete your booking. They will, however, have already asked for your permission to share information with the service.

More information can be found on Google's Privacy Policy.

Chapter 15

Device Support

This chapter deals with troubleshooting your Google Home device and the support that is available to you.

How to FDR (Factory Data Reset) Your Google Home Device

This procedure will restore Google Home to the default factory settings. It will clear all of your data and settings and this cannot be reversed. Once a factory reset has been done, you will need to start from the beginning and set up your Google Home as a new device as well as linking any devices that were previously linked, such as home control, speakers, TVs, etc.

Google Home Device:

1. Press the microphone mute button on the back of the Google Home device and hold it for about 15 seconds. Your Google Assistant will verbally confirm that Google Home is being reset.

You cannot do a Factory Data Reset using voice control or through the Google Home app.

How to Reboot Google Home Device

You may find that a reboot can fix a number of issues on your Google Home device and there are two ways to do it:

Google Home App:

1. Ensure that your smartphone or tablet and your Google Home device are connected to the same Wi-Fi network
2. Open Google Home app
3. Tap on **Devices** – top right corner of the screen – and you will see a list of the available Google Home devices
4. Scroll down and locate the device card for the device you want to be rebooted
5. Tap on **Menu** – top right corner of the card
6. Tap on **Settings**
7. Now tap on **More** - top right corner of the card
8. Tap on **Reboot**

Disconnecting the Google Home Device:

1. Remove the power cord from the electrical outlet and from the back of the Google Home device
2. Leave Google Home device unplugged for at least a minute
3. Reconnect the power cord to the device and to the electrical outlet

Please note that rebooting your Google Home device is not the same as carrying out a Factory Data Reset. You will not lose all of your data or settings by rebooting your device

Finding Answers About Google Home

Connectivity Issues

If you lose your Wi-Fi connection or you have changed your Wi-Fi access details (username and/or password) recently, you might find that you have to set your Google Home device up again. Follow these steps to try fixing the issues:

1. Open the Google Home app
2. Tap on **Devices** - top right corner of the screen – to see all of your available devices
3. Scroll down to find the relevant device card for Google Home device
4. Tap on **Set Up** and follow the steps needed to set up your device again.

If you need to change the Wi-Fi network that you have connected to your Google Home device, follow these steps:

Android Mobile Device:

1. Open Google Home app
2. Tap on **Menu>Devices** and find the device card for the Google Home device
3. Using the device card, tap on **Menu>Settings>Wi-Fi**
4. The Wi-Fi network you see is the one you are connected to
5. To change the network, go down the list and find the network that you want to connect to
6. Tap on it and type in the password to the new network

7. Tap on **Ok** and wait for the device to connect to the network

iOS Mobile Device

1. Open Google Home app
2. Tap on **Menu>Devices** and locate the device card for the Google Home device
3. On that card, tap on **Menu>Settings>Wi-Fi**
4. The network that you see is the one you are connected to
5. To change to a new Wi-Fi connection tap on **Forget this Network**
6. Tap on **Forget Wi-Fi Network**
7. Now you will need to set up your iOS device again with the new network

Getting Google Home Updates

To make the most out of your Google Home device and to take advantage of all the latest features and security patches, your device will need to be updated on a regular basis. Google will issue software updates on a regular basis and your Google Home device will download these updates automatically – you should not need to anything to update your device.

While your Google Home device is updating, you will not be able to ask Google Home any questions or ask for help completing any tasks. You must wait until the update has finished before you connect with Google Assistant again. Here is what to expect while your Google Home device is updating:

- Your Google Home app will indicate the progress of the update

- The LED lights on top of your Google Home device will indicate that Google Home is being updated

- Updates usually take around 10 minutes. If it takes too long, check the LED lights to make sure that Google Home is still updating. You should see a white spinning light

- If necessary, you can reboot your Google Home device by removing the plug, waiting for a minute and then reconnecting it. The update will automatically be restarted

- If the issue carries on, you may need to consider a Factory Data Reset and attempt to set up your Google Home device again. Try this on a 2.4 GHz Wi-Fi network if you can

The reason for that is this – there is an optional feature in 802.11AC called STBC – Space-time block coding. This lets a transmitter transmit a number of copies of streams of data via several antennas. It will also let a receiver chose the best of those data copies so that reliability is improved. There is a known issue between Google Home and certain 802.11AC router models that try to use the STBC feature on a 5 GHz network, which may result in a slower download of the update when it is done Over the Air (OTA).

The following routers are those that are known to have these issues. If yours is one of these, you will need to do a Factory Data Reset and then set up your Google Home device on a 2.4 GHz router. Once you have completed the setup and the updates have been successful you can go back to your 5 GHz network through the Google Home app:

- Buffalo WZR_D1800H
- Asus RT_AC66U
- Asus RT_AC68W
- Netgear R6100
- Netgear R6200
- Netgear R6300
- TPLink Archer C7 v1
- TPLink Archer C7 v2

Conclusion

Thank you again for purchasing this book!

I hope this book was able to help you to understand how to set up and use your new Google Home personal assistant.

The next step is to play! Have some fun with Google Home and continue your learning experience. Stretch your imagination and see just how far you can go with Google Home and, above all, have fun and enjoy!

Finally, if you enjoyed this book, then I'd like to ask you for a favor, would you be kind enough to leave a review for this book on amazon? It'd be greatly appreciated!
Thank you and good luck

CPSIA information can be obtained
at www.ICGtesting.com
Printed in the USA
LVOW13s1623270717
542867LV00012B/986/P